Collins

Maths
in 5 minutes

Quick practice activities

Chief editor: Zhou Jieying
Consultant: Fan Lianghuo

CONTENTS

HOW TO USE THIS BOOK

The best way to help your child to build their confidence in maths and improve their number skills is to give them lots and lots of practice in the key facts and skills.

Written by maths experts, this series will help your child to become fluent in number facts, and help them to recall them quickly – both are essential for succeeding in maths.

This book provides ready-to-practise questions that comprehensively cover the number curriculum for Year 5. It contains 40 topic-based tests, each 5 minutes long, to help your child build up their mathematical fluency day-by-day.

Each test is divided into three Steps:

- **Step 1: Warm-up (1 minute)**
  This exercise helps your child to revise maths they should already know and gives them preparation for Step 2.
- **Step 2: Rapid calculation ($2\frac{1}{2}$ minutes)**
  This exercise is a set of questions focused on the topic area being tested.
- **Step 3: Challenge ($1\frac{1}{2}$ minutes)**
  This is a more testing exercise designed to stretch your child's mental abilities.

Some of the tests also include:

- a Tip to help your child answer questions of a particular type.
- a Mind Gym puzzle – this is a further test of mental agility and is not included in the 5-minute time allocation.

Your child should attempt to answer as many questions as possible in the time allowed at each Step. Answers are provided at the back of the book.

To help to measure progress, each test includes boxes for recording the date of the test, the total score obtained and the total time taken.

ACKNOWLEDGEMENTS

The authors and publisher are grateful to the copyright holders for permission to use quoted materials and images.

All images are © HarperCollins*Publishers* Ltd and © Shutterstock.com

Every effort has been made to trace copyright holders and obtain their permission for the use of copyright material. The authors and publisher will gladly receive information enabling them to rectify any error or omission in subsequent editions. All facts are correct at time of going to press.

Published by Collins in association with East China Normal University Press

Collins
An imprint of HarperCollins*Publishers*
1 London Bridge Street
London SE1 9GF

HarperCollins*Publishers*
Macken House, 39/40 Mayor Street Upper,
Dublin 1, D01 C9W8, Ireland

ISBN: 978-0-00-831112-4

First published 2019

This edition published 2020

Previously published as Letts

10 9 8 7 6 5 4 3

British Library Cataloguing in Publication Data.

A CIP record of this book is available from the British Library.

Publisher: Fiona McGlade
Consultant: Fan Lianghuo
Authors: Zhou Jieying, Chen Weihua and Xu Jing
Editors: Ni Ming and Xu Huiping
Contributor: Paul Hodge
Project Management and Editorial: Richard Toms, Lauren Murray and Marie Taylor
Cover Design: Sarah Duxbury and Kevin Robbins
Inside Concept Design: Paul Oates and Ian Wrigley
Layout: Jouve India Private Limited

Printed in Great Britain by Ashford Colour Press Ltd.

MIX
Paper | Supporting responsible forestry
FSC™ C007454

This book is produced from independently certified FSC™ paper to ensure responsible forest management.

For more information visit: www.harpercollins.co.uk/green

Date: _____

Day of Week: _____

STEP 1 (1 min) Warm-up

Start the timer

Fill in the missing numbers.

$9708 = \boxed{} \times 1000 + \boxed{} \times 100 + \boxed{} \times 10 + \boxed{} \times 1$

$13\,904 = \boxed{} \times 10\,000 + \boxed{} \times 1000 + \boxed{} \times 100 + \boxed{} \times 10 + \boxed{} \times 1$

$4250 = \boxed{} \times \boxed{} + \boxed{} \times \boxed{} + \boxed{} \times \boxed{} + \boxed{} \times \boxed{}$

$37\,521 = \boxed{} \times \boxed{} + \boxed{} \times \boxed{} + \boxed{} \times \boxed{} + \boxed{} \times \boxed{} + \boxed{} \times \boxed{}$

$3549 = \boxed{} \times \boxed{} + \boxed{} \times \boxed{} + \boxed{} \times \boxed{} + \boxed{} \times \boxed{}$

STEP 2 (2.5 min) Rapid calculation

Start the timer

Complete the table.

Description of number	Number in digits
5 ten thousands, 2 hundreds and 1 ones	
3 thousands, 2 hundreds, 9 tens and 2 ones	
45 ten thousands and 7 thousands	
8 ten millions, 4 ten thousands, 6 hundreds and 3 tens	
4 millions, 3 ten thousands, 7 hundreds and 6 tens	
5 millions and 3 thousands	
9 ten millions, 6 millions, 2 hundred thousands and 5 thousands	
7 millions, 9 hundred thousands and 8 hundreds	

STEP 3 (1.5 min) Challenge

Start the timer

Complete the table.

Description of number	Number in digits	Read as
20 millions and 402 thousands		
3 millions, 50 thousands and 6 ones		
500 millions, 500 thousands and 5 hundreds		

Time spent: _____ min _____ sec. Total: _____ out of 19

©HarperCollins*Publishers* 2019

STEP 1 (1 min) Warm-up

Start the timer

Complete the table.

Written as	Read as
	four thousand, eight hundred and seventy-two
	six thousand, four hundred and eight
	nine thousand and one
	three thousand and twenty-nine
	seven thousand, seven hundred and seventy
	five thousand, seven hundred

STEP 2 (2.5 min) Rapid calculation

Start the timer

Complete the table.

Written as	Read as
	seven thousand, two hundred
	four million, nine hundred and twenty-seven thousand, and five
	twenty-four million, six thousand
	thirty-six million, four hundred and twenty thousand
	fifty million, eighty thousand and twenty-four
	two hundred and sixty million
	eight billion and three million
	one billion, five hundred and four million and eighty thousand
	ninety-five million and twenty thousand
	one hundred and six million, six hundred and sixty

STEP 3 (1.5 min) Challenge

Start the timer

Fill in each box with >, < or =.

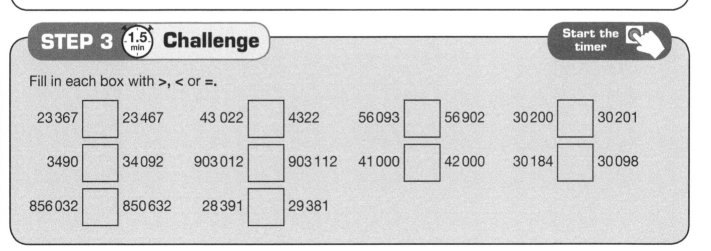

23 367 ☐ 23 467   43 022 ☐ 4322   56 093 ☐ 56 902   30 200 ☐ 30 201

3490 ☐ 34 092   903 012 ☐ 903 112   41 000 ☐ 42 000   30 184 ☐ 30 098

856 032 ☐ 850 632   28 391 ☐ 29 381

Date: _____

Day of Week: _____

STEP 1 (1 min) Warm-up

Start the timer

Write the next four numbers in each sequence.

Sequence					
+ 100	5645				
− 100	7076				
+ 1000	45 863				
− 1000	93 207				

STEP 2 (2.5 min) Rapid calculation

Start the timer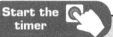

Write the next four numbers in each sequence.

Sequence					
+ 1000	78 346				
− 1000	95 079				
+ 10 000	467 374				
− 10 000	230 458				
+ 100 000	545 983				
− 100 000	909 921				

STEP 3 (1.5 min) Challenge

Start the timer

Identify and complete each sequence.

Sequence					
	483 473	473 473	463 473		
	29 044	28 944	28 844		
	678 547	578 547	478 547		
	358 975	359 975	360 975		

Time spent: _____ min _____ sec. Total: _____ out of 52

Date: _____

Day of Week: _____

STEP 1 (1 min) Warm-up

Start the timer

Answer these.

450 + 48 = ☐   800 − 360 = ☐   32 × 25 = ☐

2400 ÷ 20 = ☐   24 × 125 = ☐   1200 ÷ 50 = ☐

128 + 52 = ☐   504 − 497 = ☐

STEP 2 (2.5 min) Rapid calculation

Start the timer

 Rounding: If the first of the digits to be removed is 4 or less, that digit and all digits to its right become zero (rounding down). Leave the last remaining digit the same. If the first of the digits to be removed is 5 or greater, increase the last remaining digit by 1 and all digits to its right become zero (rounding up).

Round the numbers to the nearest ten thousand.

302 196 ≈ ☐   52 830 ≈ ☐   59 940 678 ≈ ☐

56 706 329 ≈ ☐   1 502 496 ≈ ☐   819 179 ≈ ☐

3 504 177 ≈ ☐   712 196 ≈ ☐   898 137 ≈ ☐

56 836 417 ≈ ☐   738 250 ≈ ☐   444 372 ≈ ☐

7 408 497 ≈ ☐   3 963 508 ≈ ☐   50 039 979 ≈ ☐

STEP 3 (1.5 min) Challenge

Start the timer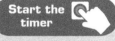

Round the numbers in the table below to the nearest ten thousand and hundred thousand.

	Ten thousand	Hundred thousand
4 503 700		
7 372 107		
7 949 270		
3 185 804		

Time spent: _____ min _____ sec. Total: _____ out of 31

Roman numerals to 1000

Date: _____

Day of Week: _____

STEP 1 (1 min) Warm-up

 Start the timer

Answer these.

I = ☐   V = ☐   XX = ☐   X = ☐

C = ☐   IV = ☐   VI = ☐   M = ☐

L = ☐   D = ☐

STEP 2 (2.5 min) Rapid calculation

 Start the timer

Answer these.

III = ☐   XIV = ☐   CCC = ☐

XCVIII = ☐   XXIX = ☐   VII = ☐

XIX = ☐   DC = ☐   MDC = ☐

MMX = ☐   IX = ☐   CL = ☐

CM = ☐   LXV = ☐   XLIX = ☐

STEP 3 (1.5 min) Challenge

 Start the timer

Answer these.

XXXIV = ☐   XCV = ☐   LXXX = ☐

MC = ☐   MD = ☐   MCM = ☐

MCMLXX = ☐   MCMXCV = ☐

Time spent: _____ min _____ sec. Total: _____ out of 33

©HarperCollinsPublishers 2019

Date: _____

Day of Week: _____

STEP 1 (1 min) Warm-up

Start the timer

Fill in the missing words.

1. Numbers with "+", such as +23 and +47, are .... and numbers with "–", such as

–16 and –23, are ....

2. The "+" sign in front of a positive number .... be omitted and the "–" sign in

front of a negative number .... be omitted.

3. 0 is neither a .... number nor a .... number.

4. In daily life, we often use .... numbers and .... numbers to

represent the idea of opposite quantities.

STEP 2 (2.5 min) Rapid calculation

Start the timer

Place these numbers into the correct set.

-19 $+24$ -2.08 5.9 0 $+10.6$ $-\dfrac{2}{25}$ 32.5 -57 $+\dfrac{4}{7}$

Positive numbers   **Negative numbers**

STEP 3 (1.5 min) Challenge

Start the timer

A warehouse sets 100 kg as a baseline to record the mass of boxes of fruit. A "+" sign is used where boxes weigh more than 100 kg and a "–" sign is used where boxes weigh less than 100 kg. Fill in the actual mass of each box.

	Box 1	Box 2	Box 3	Box 4	Box 5	Box 6	Box 7
Recorded mass (kg)	+7	–4	+6	–2.4	+3.5	+6.3	–5.8
Actual mass (kg)							

Time spent: _____ min _____ sec. Total: _____ out of 25

Date: _____

Day of Week: _____

STEP 1 (1 min) Warm-up

Start the timer

Choose the correct words from these options to complete each sentence.

positive direction   **unit length**   **origin**   **left**   **positive**   **right**

On a number line, the point representing 0 is called the ....

All points representing positive numbers are on the .... side of the origin and all points

representing negative numbers are on the .... side of the origin.

STEP 2 (2.5 min) Rapid calculation

Start the timer

1. Which number does each point represent?

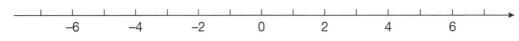

$A =$ ☐ $B =$ ☐ $C =$ ☐

$D =$ ☐ $E =$ ☐

2. Mark +5, −1, 0, 3.5, $-6\frac{1}{2}$ and $+\frac{3}{4}$ on this number line.

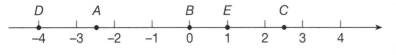

3. The point representing +6 is on the .... side of the origin and is ☐ units from the origin.

4. The point representing ☐ is on the left side of the origin and is 3.5 units from the origin.

STEP 3 (1.5 min) Challenge

Start the timer

1. On the .... side of the origin, the point ☐ units from the origin represents −5.5.

2. If the distance from a point to the origin is seven units, the number that the point represents is ☐.

3. The point that is three units from the point −1 is ☐.

4. Between +4.5 and −4.5, there are ☐ integers. Of these, ☐ are positive.

5. A point moves from −1 three units to the left, and then six units to the right. It is now at ☐.

6. The integers that are less than two units away from the point 2 are ☐, ☐ and ☐.

Time spent: _____ min _____ sec. Total: _____ out of 27

Date: _____

Day of Week: _____

STEP 1 (1 min) **Warm-up**

Start the timer

Answer these.

```
    7  1  9  8          9  2  8  3          9  0  0  9          2  3  6  8
 +  8  6  6  9       +  1  1  9  3       +  5  8  4  9       +  4  8  8  3
 _____         _____         _____         _____

 _____         _____         _____         _____
```

STEP 2 (2.5 min) **Rapid calculation**

Start the timer

Answer these.

```
    3  2  4  8  5        6  8  5  3  7        6  6  4  6  1        4  6  5  1  5
 +  2  1  6  2  2     +  6  3  7  3  3     +  1  8  6  3  4     +  6  9  9  1  8
 _____      _____      _____      _____

 _____      _____      _____      _____


    1  4  0  3  0        9  6  4  6  8        4  6  8  5  3        9  4  4  8  9
 +  1  2  8  4  9     +  5  5  2  3  1     +  9  0  5  5  5     +  7  3  8  7  2
 _____      _____      _____      _____

 _____      _____      _____      _____
```

STEP 3 (1.5 min) **Challenge**

Start the timer

Answer these.

```
    7  5  7  7  0              9  9  7  1  7
    1  9  3  0  2              4  5  3  0  7
 +  3  7  0  2  9           +  9  9  3  1  0
 _____            _____

 _____            _____


    4  1  4  2  6              8  6  6  2  0
    2  5  1  4  0              5  0  9  5  0
 +  9  0  8  5  3           +  9  2  7  5  3
 _____            _____

 _____            _____
```

Date: _____

Day of Week: _____

Start the timer

Answer these.

```
    7 7 7 1        6 2 0 9        5 3 5 0        8 7 6 9
  - 6 5 4 3      - 2 3 8 7      - 2 2 9 5      - 7 3 5 9
  ─────────      ─────────      ─────────      ─────────

  ─────────      ─────────      ─────────      ─────────
```

Start the timer

Answer these.

```
    7 5 6 0 9          5 1 2 1 6          9 4 5 0 2
  - 7 3 7 0 6        - 2 6 9 2 8        - 6 6 8 3 3
  ───────────        ───────────        ───────────

  ───────────        ───────────        ───────────


    9 4 0 9 5          6 2 9 4 6          5 9 5 8 0
  - 6 3 0 5 3        - 2 4 9 8 8        - 4 2 1 5 2
  ───────────        ───────────        ───────────

  ───────────        ───────────        ───────────
```

Start the timer

Fill in the missing numbers..

```
  ☐ 4 9 2 ☐        5 9 ☐ 7 8        8 4 ☐ 8 4
- 5 9 ☐ 0 9      - ☐ 7 0 6 ☐      - ☐ 4 2 3 ☐
───────────      ───────────      ───────────
  3 5 6 1 9        3 2 0 1 5        7 0 7 5 4


  ☐ 2 4 5 ☐        5 1 ☐ 2 5        ☐ 8 2 0 ☐
- 2 8 ☐ 5 4      - ☐ 5 6 5 ☐      - 2 0 ☐ 7 7
───────────      ───────────      ───────────
  3 3 7 0 4        5 3 6 8          7 2 3 0
```

Time spent: _____ min _____ sec. Total: _____ out of 16

©HarperCollins*Publishers* 2019

Date: _____

Day of Week: _____

STEP 1 (1 min) Warm-up

Start the timer

 TIP *Multiplying a number by …*

… 10 means moving the digits one place to the left

… 100 means moving the digits two places to the left

… 1000 means moving the digits three places to the left.

Answer these.

2.3 × 10 = ☐   0.25 × 10 = ☐   1.93 × 10 = ☐   1.15 × 10 = ☐

9.6 × 10 = ☐   100 ÷ 10 = ☐   17 ÷ 10 = ☐   2.4 ÷ 10 = ☐

18.2 ÷ 10 = ☐   41.9 ÷ 10 = ☐

STEP 2 (2.5 min) Rapid calculation

Start the timer

Answer these.

0.81 × 10 = ☐   3.847 × 10 = ☐   9.23 × 10 = ☐   3.6 × 10 = ☐

780 × 10 = ☐   86 ÷ 10 = ☐   12.93 ÷ 10 = ☐   4.6 ÷ 10 = ☐

19.3 ÷ 10 = ☐   0.99 ÷ 10 = ☐   7.956 × 100 = ☐   17.5 ÷ 100 = ☐

423 ÷ 100 = ☐   0.35 × 1000 = ☐   90 ÷ 100 = ☐   356 ÷ 1000 = ☐

STEP 3 (1.5 min) Challenge

Start the timer

Answer these.

7.6 ÷ 10 ÷ 10 = ☐   3.8 × 10 × 10 = ☐   55 ÷ 10 ÷ 10 = ☐

0.99 × 10 × 10 = ☐   34.5 × 10 ÷ 100 = ☐   26.7 ÷ 100 × 10 = ☐

6 ÷ 100 × 10 = ☐   3.569 × 100 ÷ 10 × 100 = ☐

Time spent: _____ min _____ sec. Total: _____ out of 34

Date: _____

Day of Week: _____

STEP 1 (1 min) **Warm-up**

Start the timer

Answer these.

$5.6 \times 100 =$ ☐

$3.5 \times 10 =$ ☐

$0.234 \times 100 =$ ☐

$0.2 \times 100 =$ ☐

$8.02 \times 10 =$ ☐

$39 \div 10 =$ ☐

$2.9 \div 100 =$ ☐

$0.41 \times 10 =$ ☐

$0.025 \div 10 =$ ☐

$1.06 \div 100 =$ ☐

STEP 2 (2.5 min) **Rapid calculation**

Start the timer

Answer these.

$76.8 \times 10 =$ ☐

$0.629 \times 100 =$ ☐

$25.9 \div 1000 =$ ☐

$3.18 \times 1000 =$ ☐

$470.8 \div 10 =$ ☐

$0.008 \times 100 =$ ☐

$12.16 \div 100 =$ ☐

$0.29 \times 1000 =$ ☐

$26.01 \div 10 =$ ☐

$30.5 \times 100 =$ ☐

$7.04 \div 100 =$ ☐

$3.28 \times 1000 =$ ☐

$8.4 \div 100 =$ ☐

$0.059 \times 10 =$ ☐

$4.4 \div 100 =$ ☐

STEP 3 (1.5 min) **Challenge**

Start the timer

Fill in the missing numbers.

$3.4 \times$ ☐ $= 340$

$5.14 \times$ ☐ $= 51.4$

$93.7 \times$ ☐ $= 9370$

$4.06 \times$ ☐ $= 4060$

$59 \div$ ☐ $= 0.59$

$7.8 \div$ ☐ $= 0.78$

$480 \div$ ☐ $= 0.48$

$0.02 \div$ ☐ $= 0.0002$

Time spent: _____ min _____ sec. Total: _____ out of 33

Date: _____

Day of Week: _____

STEP 1 (1 min) Warm-up

 Start the timer

Answer these.

24 × 50 = ☐　　14 × 50 = ☐　　16 × 30 = ☐　　19 × 40 = ☐

150 × 6 = ☐　　12 × 80 = ☐　　13 × 70 = ☐　　50 × 40 = ☐

STEP 2 (2.5 min) Rapid calculation

 Start the timer

Answer these.

20 × 17 × 5 = ☐　　25 × 15 × 4 = ☐　　125 × 9 × 8 = ☐

5 × 37 × 8 = ☐　　86 + 14 × 50 = ☐　　15 × 33 × 2 = ☐

150 × 6 × 4 = ☐　　20 × 20 × 5 = ☐　　2400 − 30 × 40 = ☐

100 + 20 × 45 = ☐　　30 × 30 + 30 = ☐　　50 × 5 × 3 = ☐

190 × 30 − 1900 = ☐　　66 × 4 × 25 = ☐　　85 + 23 × 30 = ☐

STEP 3 (1.5 min) Challenge

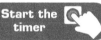 Start the timer

Answer these.

33 × 28 = ☐　　　　22 × 19 = ☐

24 × 26 = ☐　　　　87 × 11 = ☐

32 × 15 = ☐　　　　22 × 18 = ☐

65 × 65 = ☐　　　　25 × 11 = ☐

Time spent: _____ min _____ sec. Total: _____ out of 31

13 Dividing two- or three-digit numbers by a two-digit number

Date: _____

Day of Week: _____

STEP 1 (1 min) Warm-up

Start the timer

Answer these.

760 ÷ 40 = ☐   780 ÷ 60 = ☐   550 ÷ 50 = ☐   720 ÷ 60 = ☐

870 ÷ 30 = ☐   500 ÷ 50 = ☐   840 ÷ 30 = ☐   920 ÷ 40 = ☐

600 ÷ 50 = ☐   700 ÷ 50 = ☐

STEP 2 (2.5 min) Rapid calculation

Start the timer

Answer these.

56 ÷ 14 = ☐   65 ÷ 13 = ☐   84 ÷ 28 = ☐

51 ÷ 17 = ☐   950 ÷ 19 = ☐   800 ÷ 16 = ☐

540 ÷ 18 = ☐   760 ÷ 19 = ☐   92 ÷ 23 = ☐

850 ÷ 17 = ☐   63 ÷ 21 = ☐   840 ÷ 14 = ☐

900 ÷ 15 = ☐   680 ÷ 17 = ☐   108 ÷ 36 = ☐

STEP 3 (1.5 min) Challenge

Start the timer

Answer these.

(45 ÷ 15) × 30 = ☐   (17 × 60) ÷ 51 = ☐   (120 ÷ 20) × 24 = ☐

(960 ÷ 16) × 12 = ☐   (620 × 2) ÷ 40 = ☐   680 ÷ 17 × 25 = ☐

(175 ÷ 25) × 30 = ☐   570 ÷ (68 − 49) = ☐

(38 ÷ 19) × 28 = ☐   (770 ÷ 11) × 18 = ☐

Time spent: _____ min _____ sec. Total: _____ out of 35

Date: _____

Day of Week: _____

STEP 1 (1 min) Warm-up

Start the timer

Answer these.

1. $15 \div 3 =$ ☐

2. $18 \div 9 =$ ☐

3. $45 \div 5 =$ ☐

4. $72 \div 8 =$ ☐

$150 \div 30 =$ ☐

$180 \div 90 =$ ☐

$450 \div 50 =$ ☐

$720 \div 80 =$ ☐

$1500 \div 30 =$ ☐

$1800 \div 90 =$ ☐

$4500 \div 50 =$ ☐

$7200 \div 80 =$ ☐

STEP 2 (2.5 min) Rapid calculation

Start the timer

Answer these.

$9600 \div 24 =$ ☐

$1440 \div 24 =$ ☐

$1280 \div 16 =$ ☐

$8400 \div 28 =$ ☐

$1440 \div 16 =$ ☐

$1380 \div 23 =$ ☐

$1040 \div 26 =$ ☐

$1350 \div 27 =$ ☐

$1300 \div 26 =$ ☐

$1400 \div 28 =$ ☐

$1750 \div 25 =$ ☐

$1680 \div 28 =$ ☐

$1470 \div 21 =$ ☐

$1050 \div 15 =$ ☐

$1200 \div 24 =$ ☐

STEP 3 (1.5 min) Challenge

Start the timer

Answer these.

$(46 \times 20) \div 23 =$ ☐

$(45 \times 50) \div 15 =$ ☐

$(620 \div 31) \times 80 =$ ☐

$(99 \div 11) \times 15 =$ ☐

$(27 \times 110) \div 30 =$ ☐

$39 \times (380 \div 19) =$ ☐

$41 \times (250 \div 50) =$ ☐

$(320 \times 15) \div 40 =$ ☐

$(36 \times 40) \div 24 =$ ☐

$(84 \div 21) \times 32 =$ ☐

Date: _____

Day of Week: _____

STEP 1 (1 min) Warm-up

Start the timer

Answer these.

60 ÷ 4 = ☐   128 ÷ 8 = ☐   84 ÷ 7 = ☐   120 ÷ 8 = ☐

54 ÷ 3 = ☐   126 ÷ 6 = ☐   39 ÷ 3 = ☐   90 ÷ 5 = ☐

96 ÷ 6 = ☐   32 ÷ 2 = ☐

STEP 2 (2.5 min) Rapid calculation

Start the timer

Answer these.

80 ÷ 5 = ☐   98 ÷ 7 = ☐   78 ÷ 6 = ☐

68 ÷ 4 = ☐   108 ÷ 6 = ☐   48 ÷ 4 = ☐

94 ÷ 2 = ☐   63 ÷ 3 = ☐   125 ÷ 5 = ☐

72 ÷ 18 = ☐   75 ÷ 15 = ☐   64 ÷ 4 = ☐

95 ÷ 5 = ☐   350 ÷ 10 = ☐   81 ÷ 3 = ☐

STEP 3 (1.5 min) Challenge

Start the timer

Answer these.

256 ÷ 8 = ☐   732 ÷ 4 = ☐

156 ÷ 4 = ☐   456 ÷ 3 = ☐

396 ÷ 4 = ☐   270 ÷ 5 = ☐

297 ÷ 9 = ☐   104 ÷ 13 = ☐

Time spent: _____ min _____ sec. Total: _____ out of 35

Date: _____

Day of Week: _____

STEP 1 (1 min) Warm-up

Start the timer

1. The factors of 16 are ....

2. The factors of 24 are ....

3. The factors of 98 are ....

4. The first three multiples of 32 are ....

5. The first five multiples of 18 are ....

STEP 2 (2.5 min) Rapid calculation

Start the timer

1. The common factors of 16 and 24 are ....

2. The common factors of 15 and 25 are ....

3. The common factors of 27 and 18 are ....

4. The common factors of 36 and 32 are ....

5. Five common multiples of 6 and 4 are ....

6. Five common multiples of 12 and 8 are ....

7. Five common multiples of 5 and 15 are ....

8. Five common multiples of 6 and 9 are ....

STEP 3 (1.5 min) Challenge

Start the timer

1. The greatest common factor of:

16 and 4 is ☐   18 and 12 is ☐

12 and 24 is ☐   16 and 64 is ☐

2. The least common multiple of:

16 and 8 is ☐   7 and 13 is ☐

8 and 12 is ☐   12 and 9 is ☐

Time spent: _____ min _____ sec. Total: _____ out of 21

Date: _____

Day of Week: _____

STEP 1 (1 min) Warm-up

 Start the timer

Answer these.

$9^2 = $ ☐ $5^2 = $ ☐ $3^2 = $ ☐ $4^2 = $ ☐

$8^2 = $ ☐ $7^3 = $ ☐ $4^3 = $ ☐ $1^3 = $ ☐

$2^3 = $ ☐ $3^3 = $ ☐

STEP 2 (2.5 min) Rapid calculation

 Start the timer

Answer these.

$11^2 = $ ☐ $5^3 = $ ☐ $13^2 = $ ☐

$6^3 = $ ☐ $18^2 = $ ☐ $10^3 = $ ☐

$12^2 = $ ☐ $8^3 = $ ☐ $15^2 = $ ☐

$8 = $ ☐^{3} $19^2 = $ ☐ $196 = $ ☐^{2}

$20^2 = $ ☐ $64 = $ ☐^{2} $16^2 = $ ☐

STEP 3 (1.5 min) Challenge

 Start the timer

Answer these.

$4^3 = $ ☐^{2} $1^2 = $ ☐^{3}

$169 = $ ☐^{2} $121 = $ ☐^{2}

$343 = $ ☐^{3} $225 = $ ☐$^2 \times $ ☐^{2}

$196 = $ ☐$^2 \times $ ☐^{2} $324 = $ ☐$^2 \times $ ☐^{2}

Time spent: _____ min _____ sec. Total: _____ out of 33

Date: _____

Day of Week: _____

STEP 1 (1 min) **Warm-up**

Start the timer

Look at the numbers in the box:

| 72 | 48 | 30 | 62 | 15 | 80 | 45 | 60 | 58 |

1. Write down the numbers that are divisible by 2. ....

2. Write down the numbers that are divisible by 5. ....

STEP 2 (2.5 min) **Rapid calculation**

Start the timer

Answer these.

$450 \div 5 =$ ☐ $390 \div 5 =$ ☐ $250 \div 5 =$ ☐ $430 \div 5 =$ ☐

$450 \div 2 =$ ☐ $390 \div 2 =$ ☐ $250 \div 2 =$ ☐ $430 \div 2 =$ ☐

$940 \div 5 =$ ☐ $820 \div 5 =$ ☐ $760 \div 5 =$ ☐ $580 \div 5 =$ ☐

$940 \div 2 =$ ☐ $820 \div 2 =$ ☐ $760 \div 2 =$ ☐ $580 \div 2 =$ ☐

STEP 3 (1.5 min) **Challenge**

Start the timer

Answer these.

$185 \div 5 =$ ☐ $770 \div 5 =$ ☐

$840 \div 2 =$ ☐ $486 \div 2 =$ ☐

$290 \div 5 =$ ☐ $955 \div 5 =$ ☐

$658 \div 2 =$ ☐ $664 \div 2 =$ ☐

$745 \div 5 =$ ☐ $398 \div 2 =$ ☐

Time spent: _____ min _____ sec. Total: _____ out of 38

Date: _____

Day of Week: _____

STEP 1 (1 min) Warm-up

Start the timer

(TIP) *A natural number (a positive whole number) can be classified according to its factors.*

*A **prime number** has only two factors, 1 and itself.*

*A **composite number** has three or more factors.*

Since 1 has only one factor, it is neither composite nor prime.

Look at the numbers in the box:

| 72 | 9 | 2 | 15 | 23 | 49 | 31 | 97 | 56 | 67 | 63 |

1. Write down the prime numbers. ....

2. Write down the composite numbers. ....

STEP 2 (2.5 min) Rapid calculation

Start the timer

Write down the prime factorisation of each number.

45 = ...   32 = ...   66 = ...

27 = ...   48 = ...   81 = ...

54 = ...   84 = ...   96 = ...

STEP 3 (1.5 min) Challenge

Start the timer

Write down the prime factorisation of each number.

68 = ...   92 = ...

88 = ...   78 = ...

72 = ...   85 = ...

30 = ...   69 = ...

Time spent: _____ min _____ sec. Total: _____ out of 28

Date: _____

Day of Week: _____

STEP 1 (1 min) **Warm-up** — Start the timer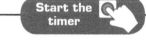

Answer these.

$45 + 54 =$ ☐ $\qquad$ $300 \div 25 =$ ☐ $\qquad$ $235 - 98 =$ ☐

$305 \times 8 =$ ☐ $\qquad$ $(320 \div 4) \times 25 =$ ☐ $\qquad$ $(15 \times 8) + 12 =$ ☐

$(31 \times 11) - 11 =$ ☐ $\qquad$ $57 \times 2 \times 5 =$ ☐ $\qquad$ $(26 \times 49) + 26 =$ ☐

STEP 2 (2.5 min) **Rapid calculation** — Start the timer

Write down one number sentence for each set of calculations. The first one has been done for you.

1. $18 \times 4 = 72$, $72 - 12 = 60$, $60 \div 6 = 10$ $\qquad$ **$(18 \times 4 - 12) \div 6 = 10$**

2. $470 - 362 = 108$, $356 \div 4 = 89$, $89 \times 108 = 9612$ $\qquad$ ....

3. $238 - 195 = 43$, $43 \times 9 = 387$, $3870 \div 387 = 10$ $\qquad$ ....

4. $18 + 35 = 53$, $1360 - 247 = 1113$, $1113 \div 53 = 21$ $\qquad$ ....

5. $49 + 33 = 82$, $105 - 4 = 101$, $101 \times 82 = 8282$ $\qquad$ ....

6. $128 \times 15 = 1920$, $3045 - 1920 = 1125$, $1125 \div 45 = 25$ $\qquad$ ....

STEP 3 (1.5 min) **Challenge** — Start the timer

 (TIP) *The quotient is the result of dividing one number by another.*

Answer these.

1. The product of two 15s is subtracted from 650.

What is the result? ☐

2. 450 is divided by the sum of 50 and twice 50.

What is the quotient? ☐

3. 280 is divided by the sum of 16 and 54.

What is the quotient? ☐

4. The difference of 49 and 38 is multiplied by the quotient of 64 divided by 8.

What is the product? ☐

Mind Gym

The sum of five one-digit numbers is 30. The product of the same five numbers is 2520.

One of the numbers is 1 and another is 8.

Can you find the other three numbers?

☐ + ☐ + ☐ +

$1 + 8 = 30$

☐ × ☐ × ☐ ×

$1 \times 8 = 2520$

Time spent: _____ min _____ sec. Total: _____ out of 18

21 | Unit rates

Date: _____

Day of Week: _____

STEP 1 (1 min) Warm-up

Start the timer

Complete the table.

Description of rate	Rate	Unit rate
60 mm of rain over 4 days	$\dfrac{60\,mm\ of\ rain}{4\ days}$	15 mm of rain per day
6 books for £30		
120 seeds in 3 rows		
125 oranges in 25 bowls		
132 hours of work in 12 days		

STEP 2 (2.5 min) Rapid calculation

Start the timer

Complete the table.

Description of rate	Rate	Unit rate	Calculation
420 points in 6 games			[ ] points in 9 games
280 pages in 8 days			[ ] pages in 11 days
£252 for 7 hours work			£ [ ] for 15 hours work
297 girls in 9 groups			[ ] girls in 15 groups

STEP 3 (1.5 min) Challenge

Start the timer

Complete the table.

Description of rate	Rate	Unit rate	Calculation
760 kilometres on 8 litres of fuel			[ ] km on 13 L of fuel
438 points in 6 plays of a video game			876 points in [ ] plays
£756 for 9 tickets			£ [ ] for 15 tickets
720 kg for 12 crates			1080 kg for [ ] crates

Time spent: _____ min _____ sec. Total: _____ out of 32

©HarperCollins*Publishers* 2019

Date: _____

Day of Week: _____

Start the timer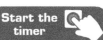

STEP 1 (1 min) Warm-up

Answer these.

702 ÷ 6 = ☐   660 + 240 = ☐   730 − 290 = ☐

2520 ÷ 7 = ☐   34 × 40 = ☐   931 − 620 = ☐

96 × 50 = ☐   75 × 101 = ☐   410 + 380 = ☐

Start the timer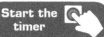

STEP 2 (2.5 min) Rapid calculation

Answer these.

325 × 5 = ☐   24 × 50 = ☐   540 − 490 = ☐

705 − 355 = ☐   101 × 17 = ☐   18 + 790 = ☐

450 ÷ 75 = ☐   300 − 102 = ☐   35 × 40 = ☐

324 ÷ 9 = ☐   180 ÷ 15 = ☐   125 × 64 = ☐

576 ÷ 6 = ☐   11 × 36 = ☐   49 × 99 = ☐

Start the timer

STEP 3 (1.5 min) Challenge

Answer these.

35 × (48 ÷ 12) = ☐   (90 × 33) ÷ 15 = ☐   (73 × 49) + 73 = ☐

320 ÷ (384 − 320) = ☐   (27 × 102) − 54 = ☐   (972 − 172) × 5 = ☐

425 − 160 − 140 = ☐   (310 + 210) ÷ 13 = ☐

Date: _____

Day of Week: _____

STEP 1 (1 min) **Warm-up**

 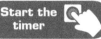

Answer these.

$\dfrac{9}{17} + \dfrac{6}{17} = \boxed{}$ $\dfrac{4}{15} + \dfrac{6}{15} = \boxed{}$ $\dfrac{8}{20} + \dfrac{9}{20} = \boxed{}$ $\dfrac{9}{16} + \dfrac{5}{16} = \boxed{}$

$\dfrac{13}{19} - \dfrac{8}{19} = \boxed{}$ $\dfrac{12}{14} - \dfrac{3}{14} = \boxed{}$ $\dfrac{15}{18} - \dfrac{7}{18} = \boxed{}$ $\dfrac{10}{11} - \dfrac{8}{11} = \boxed{}$

STEP 2 (2.5 min) **Rapid calculation**

Answer these.

$\dfrac{5}{9} + \dfrac{3}{9} = \boxed{}$ $\dfrac{48}{107} + \dfrac{36}{107} = \boxed{}$ $\dfrac{9}{11} - \dfrac{4}{11} = \boxed{}$

$\dfrac{9}{25} + \dfrac{7}{25} = \boxed{}$ $\dfrac{7}{8} - \dfrac{6}{8} = \boxed{}$ $\dfrac{56}{90} - \dfrac{37}{90} = \boxed{}$

$\dfrac{14}{46} + \dfrac{26}{46} = \boxed{}$ $\dfrac{59}{125} - \dfrac{34}{125} = \boxed{}$ $\dfrac{19}{27} + \dfrac{6}{27} = \boxed{}$

$\dfrac{17}{53} + \dfrac{21}{53} = \boxed{}$ $\dfrac{92}{217} - \dfrac{62}{217} = \boxed{}$ $\dfrac{17}{49} + \dfrac{20}{49} = \boxed{}$

$\dfrac{16}{18} - \dfrac{6}{18} = \boxed{}$ $\dfrac{35}{96} - \dfrac{23}{96} = \boxed{}$

$\dfrac{82}{113} + \dfrac{18}{113} = \boxed{}$ $\dfrac{94}{169} - \dfrac{51}{169} = \boxed{}$

STEP 3 (1.5 min) **Challenge**

Answer these.

$\dfrac{59}{100} + \dfrac{26}{100} - \dfrac{16}{100} = \boxed{}$ $\dfrac{3}{9} + \dfrac{6}{9} - \dfrac{4}{9} = \boxed{}$ $\dfrac{48}{77} - \dfrac{16}{77} - \dfrac{10}{77} = \boxed{}$

$\dfrac{159}{223} - \dfrac{59}{223} - \dfrac{91}{223} = \boxed{}$ $\dfrac{5}{17} + \dfrac{6}{17} + \dfrac{5}{17} = \boxed{}$ $\dfrac{43}{119} - \dfrac{13}{119} + \dfrac{15}{119} = \boxed{}$

$\dfrac{57}{110} - \dfrac{12}{110} + \dfrac{23}{110} = \boxed{}$ $\dfrac{35}{89} + \dfrac{34}{89} - \dfrac{15}{89} = \boxed{}$

Time spent: _____ min _____ sec. Total: _____ out of 32

Comparing fractions  24

 TIP *The more portions you divide something into, the smaller each portion will be. This means that:*
For a fraction with the numerator 1, the greater the denominator, the smaller the fraction.
If two fractions have the same denominator, the fraction with the greater numerator is greater.
If two fractions have the same numerator, the fraction with the smaller denominator is greater.

Fill in the boxes with **>, <** or **=**.

$\frac{1}{3} \square \frac{1}{5}$ $\frac{1}{10} \square \frac{1}{6}$ $\frac{1}{7} \square \frac{1}{15}$ $\frac{1}{123} \square \frac{1}{122}$

$\frac{1}{30} \square \frac{1}{50}$ $\frac{1}{4} \square \frac{1}{2}$ $\frac{1}{20} \square \frac{1}{24}$ $\frac{1}{100} \square \frac{1}{90}$

$\frac{1}{36} \square \frac{1}{63}$ $\frac{1}{45} \square \frac{1}{25}$

STEP 2 **Rapid calculation**

Fill in the boxes with **>, <** or **=**.

$\frac{5}{9} \square \frac{7}{9}$ $\frac{8}{15} \square \frac{7}{15}$ $\frac{9}{70} \square \frac{22}{70}$ $\frac{11}{70} \square \frac{11}{22}$

$\frac{3}{4} \square \frac{3}{8}$ $\frac{5}{17} \square \frac{3}{17}$ $\frac{3}{7} \square \frac{3}{17}$ $\frac{2}{9} \square \frac{2}{11}$

$\frac{4}{9} \square \frac{4}{7}$ $\frac{4}{9} \square \frac{8}{9}$ $\frac{4}{27} \square \frac{4}{72}$ $\frac{42}{97} \square \frac{24}{97}$

$\frac{1}{9} \square \frac{8}{9}$ $\frac{14}{29} \square \frac{14}{27}$ $\frac{13}{200} \square \frac{23}{200}$ $\frac{5}{69} \square \frac{5}{99}$

STEP 3 **Challenge**

Write each set of fractions in order from **greatest** to **least**.

$\frac{1}{7}$ $\frac{1}{3}$ $\frac{1}{4}$ .... $\frac{73}{99}$ $\frac{85}{99}$ $\frac{37}{99}$ .... $\frac{1}{5}$ $\frac{1}{15}$ $\frac{3}{5}$ ....

$\frac{3}{9}$ $\frac{8}{9}$ $\frac{3}{13}$ .... $\frac{13}{20}$ $\frac{4}{20}$ $\frac{13}{14}$ .... $\frac{3}{11}$ $\frac{3}{10}$ $\frac{7}{10}$ ....

Date: _____

Day of Week: _____

STEP 1 (1 min) Warm-up

Start the timer

Look at the numbers in the box:

| $\frac{7}{13}$ | $\frac{3}{8}$ | $23\frac{1}{4}$ | $\frac{9}{4}$ | $1\frac{6}{15}$ | $\frac{84}{100}$ | $40\frac{2}{3}$ | $\frac{181}{366}$ | $\frac{25}{24}$ | $\frac{19}{8}$ | $\frac{8}{7}$ | $72\frac{3}{8}$ |

1. Write down the mixed numbers. ....

2. Write down the improper fractions. ....

3. Write down the proper fractions. ....

STEP 2 (2.5 min) Rapid calculation

Start the timer

1. Write each improper fraction as a mixed number.

$$\frac{32}{7} = \boxed{} \qquad \frac{84}{9} = \boxed{} \qquad \frac{27}{4} = \boxed{}$$

$$\frac{105}{12} = \boxed{} \qquad \frac{67}{12} = \boxed{} \qquad \frac{148}{8} = \boxed{}$$

2. Write each mixed number as an improper fraction.

$$2\frac{3}{7} = \boxed{} \qquad 3\frac{1}{6} = \boxed{} \qquad 4\frac{8}{15} = \boxed{}$$

$$12\frac{7}{12} = \boxed{} \qquad 4\frac{7}{34} = \boxed{} \qquad 10\frac{9}{14} = \boxed{}$$

STEP 3 (1.5 min) Challenge

Start the timer

1. Write each mixed number as an improper fraction.

$$2\frac{1}{3} = \boxed{} \qquad 1\frac{5}{8} = \boxed{} \qquad 6\frac{7}{9} = \boxed{} \qquad 3\frac{5}{34} = \boxed{}$$

2. Write each improper fraction as a mixed number.

$$\frac{31}{7} = \boxed{} \qquad \frac{53}{8} = \boxed{} \qquad \frac{81}{9} = \boxed{} \qquad \frac{123}{12} = \boxed{}$$

Time spent: _____ min _____ sec. Total: _____ out of 32

Start the timer

STEP 1 $\binom{1}{min}$ Warm-up

Answer these.

$\dfrac{1}{5} + \dfrac{3}{5} = \boxed{}$ $\qquad$ $\dfrac{4}{9} + \dfrac{5}{9} = \boxed{}$ $\qquad$ $\dfrac{4}{15} + \dfrac{7}{15} = \boxed{}$ $\qquad$ $\dfrac{7}{24} + \dfrac{6}{24} = \boxed{}$

$\dfrac{7}{10} + \dfrac{3}{10} = \boxed{}$ $\qquad$ $\dfrac{7}{15} - \dfrac{3}{15} = \boxed{}$ $\qquad$ $\dfrac{13}{18} - \dfrac{5}{18} = \boxed{}$ $\qquad$ $\dfrac{12}{23} - \dfrac{3}{23} = \boxed{}$

$\dfrac{20}{37} - \dfrac{3}{37} = \boxed{}$ $\qquad$ $\dfrac{53}{102} - \dfrac{22}{102} = \boxed{}$

STEP 2 $\binom{2.5}{min}$ Rapid calculation

Start the timer

Answer these.

$\dfrac{1}{2} + \dfrac{1}{6} = \boxed{}$ $\qquad$ $\dfrac{3}{5} - \dfrac{2}{10} = \boxed{}$ $\qquad$ $\dfrac{4}{13} + \dfrac{5}{26} = \boxed{}$

$\dfrac{7}{10} - \dfrac{3}{20} = \boxed{}$ $\qquad$ $\dfrac{13}{23} - \dfrac{12}{46} = \boxed{}$ $\qquad$ $\dfrac{8}{15} + \dfrac{7}{45} = \boxed{}$

$\dfrac{5}{6} - \dfrac{5}{18} = \boxed{}$ $\qquad$ $\dfrac{1}{3} + \dfrac{2}{9} = \boxed{}$ $\qquad$ $\dfrac{13}{27} - \dfrac{5}{54} = \boxed{}$

$\dfrac{7}{32} + \dfrac{3}{16} = \boxed{}$ $\qquad$ $\dfrac{34}{53} + \dfrac{11}{106} = \boxed{}$ $\qquad$ $\dfrac{45}{46} - \dfrac{8}{23} = \boxed{}$

STEP 3 $\binom{1.5}{min}$ Challenge

Start the timer

Answer these.

$\dfrac{57}{70} - \dfrac{2}{35} = \boxed{}$ $\qquad\qquad$ $\dfrac{18}{75} + \dfrac{19}{25} = \boxed{}$

$\dfrac{4}{39} + \dfrac{5}{78} = \boxed{}$ $\qquad\qquad$ $\dfrac{34}{270} - \dfrac{6}{135} = \boxed{}$

$\dfrac{25}{48} - \dfrac{11}{24} = \boxed{}$ $\qquad\qquad$ $\dfrac{35}{102} + \dfrac{29}{51} = \boxed{}$

$\dfrac{37}{56} + \dfrac{13}{28} = \boxed{}$ $\qquad\qquad$ $\dfrac{67}{72} - \dfrac{17}{24} = \boxed{}$

Date: _____

Day of Week: _____

STEP 1 (1 min) Warm-up

Start the timer

Answer these.

$\frac{2}{7} + \frac{2}{7} + \frac{2}{7} = \boxed{}$

$\frac{4}{19} + \frac{4}{19} + \frac{4}{19} + \frac{4}{19} = \boxed{}$

$\frac{2}{27} + \frac{2}{27} + \frac{2}{27} + \frac{2}{27} + \frac{2}{27} = \boxed{}$

$\frac{1}{4} + \frac{1}{4} + \frac{1}{4} + \frac{1}{4} = \boxed{}$

$\frac{3}{10} + \frac{3}{10} + \frac{3}{10} = \boxed{}$

$\frac{3}{15} + \frac{3}{15} + \frac{3}{15} + \frac{3}{15} = \boxed{}$

$\frac{5}{12} + \frac{5}{12} + \frac{5}{12} + \frac{5}{12} = \boxed{}$

$\frac{1}{8} + \frac{1}{8} + \frac{1}{8} = \boxed{}$

$\frac{4}{13} + \frac{4}{13} + \frac{4}{13} + \frac{4}{13} = \boxed{}$

STEP 2 (2.5 min) Rapid calculation

Start the timer

Answer these.

$3 \times \frac{3}{13} = \boxed{}$

$\frac{3}{10} \times 2 = \boxed{}$

$5 \times \frac{1}{3} = \boxed{}$

$12 \times \frac{3}{73} = \boxed{}$

$\frac{43}{120} \times 2 = \boxed{}$

$8 \times \frac{2}{15} = \boxed{}$

$\frac{3}{14} \times 23 = \boxed{}$

$\frac{25}{100} \times 3 = \boxed{}$

$\frac{7}{10} \times 3 = \boxed{}$

$6 \times \frac{3}{5} = \boxed{}$

$4 \times 2\frac{1}{8} = \boxed{}$

$4 \times 5\frac{2}{9} = \boxed{}$

$4\frac{3}{14} \times 3 = \boxed{}$

$18\frac{2}{17} \times 6 = \boxed{}$

STEP 3 (1.5 min) Challenge

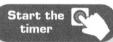
Start the timer

Answer these.

$16\frac{12}{95} \times 3 = \boxed{}$

$7 \times 3\frac{11}{100} = \boxed{}$

$7 \times 2\frac{2}{9} = \boxed{}$

$5\frac{3}{47} \times 13 = \boxed{}$

$15 \times 4\frac{2}{27} = \boxed{}$

$9 \times 12\frac{3}{28} = \boxed{}$

$4\frac{1}{12} \times 12 = \boxed{}$

$27\frac{3}{56} \times 3 = \boxed{}$

Time spent: _____ min _____ sec. Total: _____ out of 31

©HarperCollins*Publishers* 2019

Date: _____

Day of Week: _____

STEP 1 (1 min) Warm-up

Start the timer

1. Order the numbers from **least** to **greatest**.

25.068   25.860   2.468   0.864   24.68

....

2. Order the numbers from **greatest** to **least**.

5.766   6.032   0.320   5.660   5.667

....

STEP 2 (2.5 min) Rapid calculation

Start the timer

(TIP) *Inserting or removing zeroes at the end of the decimal part does not change the value of the number.*

Simplify the decimals.

0.50 = [ ]	30.2020 = [ ]	40.00 = [ ]	1.480 = [ ]
1.900 = [ ]	400.0 = [ ]	3.940 = [ ]	8.0 = [ ]
0.0810 = [ ]	7.080 = [ ]	2.910 = [ ]	51.0100 = [ ]

STEP 3 (1.5 min) Challenge

Start the timer

1. Write equivalent decimals with two decimal places.

23.7 = [ ]   67 = [ ]   12.5 = [ ]

4.5 = [ ]   87.1 = [ ]   31 = [ ]

2. Write equivalent decimals with three decimal places.

45.9 = [ ]   78.56 = [ ]   9.12 = [ ]

92 = [ ]   1.7 = [ ]   37 = [ ]

Date: _____

Day of Week: _____

STEP 1  **Warm-up**

Start the timer

Write each decimal in the correct box(es).

3.90   10.005   4.01   1000   10.001   1.4000   5.600   204.09   201.00   3.0200   50.30

Box 1	**Box 2**	**Box 3**
Value is unchanged if all zeroes are removed	Value is unchanged if all zeroes at the end are removed	No zeroes can be removed without changing the value

---

STEP 2 (2.5 min) **Rapid calculation**

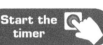 Start the timer

Simplify the decimals.

3.0400 = ⬜   7.050 = ⬜   40.030 = ⬜   8.090 = ⬜

100.20 = ⬜   9.000 = ⬜   50.5050 = ⬜   3.020 = ⬜

0.30 = ⬜   3.0300 = ⬜   6.2000 = ⬜   5.80 = ⬜

10.10 = ⬜   0.750 = ⬜   3.990 = ⬜   20.200 = ⬜

STEP 3 (1.5 min) **Challenge**

 Start the timer

Rewrite each amount in pounds. Remember to use two decimal places.

36 pence = £ ⬜   4 pence = £ ⬜

110 pence = £ ⬜   1 pound 6 pence = £ ⬜

3 pounds 4 pence = £ ⬜   90 pence = £ ⬜

46 pounds 35 pence = £ ⬜   70 pounds 3 pence = £ ⬜

Time spent: _____ min _____ sec. Total: _____ out of 38

Date: _____

Day of Week: _____

 When adding decimals, first line up the decimal points, then use column addition to add before putting a decimal point in the answer. Zeroes that do not affect the value can be omitted.

Answer these.

2.2 + 0.4 = ☐   9 + 2.3 = ☐   8.1 + 0.4 = ☐   1.3 + 10.5 = ☐

3.9 + 14 = ☐   0.7 + 1.3 = ☐   2.5 + 4.4 = ☐   2.7 + 0.8 = ☐

8.5 + 0.5 = ☐   5.2 + 0.9 = ☐

Answer these.

5.4 + 6 = ☐   0.8 + 0.92 = ☐   7.5 + 5 = ☐   1.2 + 0.5 = ☐

1.04 + 2.9 = ☐   7.2 + 2.7 = ☐   10 + 1.09 = ☐   0.37 + 2 = ☐

2.02 + 0.2 = ☐   6.6 + 0.8 = ☐   0.64 + 0.4 = ☐   8 + 2.5 = ☐

41.4 + 2.32 = ☐   3.5 + 0.9 = ☐   0.09 + 0.26 = ☐   7.5 + 0.55 = ☐

Answer these.

0.2 + 0.4 + 1.8 = ☐   3.5 + 2.7 + 6.5 = ☐

5.2 + 2.8 + 5.3 = ☐   0.6 + 1.9 + 5.4 = ☐

0.3 + 0.92 + 0.08 = ☐   4.2 + 1.9 + 4.8 = ☐

0.24 + 0.8 + 0.76 = ☐   0.8 + 2.2 + 3.6 + 1.4 = ☐

Time spent: _____ min _____ sec. Total: _____ out of 34

Date: _____

Day of Week: _____

STEP 1 (1 min) Warm-up

Start the timer

Answer these.

0.5 – 0.2 = [ ]　　2.9 – 0.3 = [ ]　　2.95 – 0.45 = [ ]　　1.8 – 0.4 = [ ]

8.24 – 0.09 = [ ]　　0.45 – 0.4 = [ ]　　1 – 0.6 = [ ]　　9.4 – 0.7 = [ ]

0.64 – 0.3 = [ ]　　4.3 – 0.5 = [ ]

STEP 2 (2.5 min) Rapid calculation

Start the timer

Answer these.

0.7 – 0.4 = [ ]　　3.8 – 1.2 = [ ]　　0.56 – 0.05 = [ ]　　0.35 – 0.25 = [ ]

8.96 – 0.6 = [ ]　　1 – 0.05 = [ ]　　4.7 – 4 = [ ]　　3.8 – 3.75 = [ ]

5 – 1.6 = [ ]　　9 – 0.09 = [ ]　　48 – 0.7 = [ ]　　2 – 0.8 = [ ]

11 – 0.2 = [ ]　　3.9 – 0.3 = [ ]　　1.07 – 0.4 = [ ]　　10.5 – 0.75 = [ ]

STEP 3 (1.5 min) Challenge

Start the timer

(TIP) *Always check if there is a more simple way to do the calculation. For example:*
7.45 – 3.2 – 1.45 = 7.45 – 1.45 – 3.2 = 6 – 3.2 = 2.8

Answer these.

8.35 – 2.35 – 0.7 = [ ]　　　　4.8 – 2.4 – 1.8 = [ ]

10 – 2.7 – 0.3 = [ ]　　　　17.3 – 0.04 – 7.3 = [ ]

20 – 0.8 – 2.2 = [ ]　　　　9.52 – 2.4 – 3.6 = [ ]

14.5 – 3.5 – 1.9 = [ ]　　　　4.56 – 0.7 – 0.56 = [ ]

Time spent: _____ min _____ sec. Total: _____ out of 34

Date: _____

Day of Week: _____

STEP 1 (1 min) **Warm-up**

Start the timer

Write each percentage as a fraction. The first one has been done for you.

$37\% = \dfrac{37}{100}$ $23\% = \dfrac{}{100}$ $51\% = \dfrac{}{100}$ $66\% = \dfrac{}{100}$

$97\% = \dfrac{}{100}$ $49\% = \dfrac{}{100}$ $88\% = \dfrac{}{100}$ $3\% = \dfrac{}{100}$

$91\% = \dfrac{}{100}$ $9\% = \dfrac{}{100}$ $24\% = \dfrac{}{100}$ $71\% = \dfrac{}{100}$

STEP 2 (2.5 min) **Rapid calculation**

Start the timer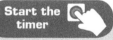

Complete the table.

Percentage	7%	77%	19%	30%	59%	90%	1%	43%	99%	5%
Fraction										
Decimal										

STEP 3 (1.5 min) **Challenge**

Start the timer

Complete the table.

Percentage	3%	62%		11%	33%		44%	85%		13%
Fraction		$\dfrac{62}{100}$	$\dfrac{80}{100}$		$\dfrac{33}{100}$	$\dfrac{8}{100}$		$\dfrac{85}{100}$	$\dfrac{60}{100}$	
Decimal	0.03		0.8	0.11		0.08	0.44		0.6	0.13

Date: _____

Day of Week: _____

STEP 1 (1 min) Warm-up

Start the timer

Convert these.

1 kg = [   ] g   4 kg = [   ] g   1000 g = [   ] kg   8000 g = [   ] kg

5000 g = [   ] kg   80 kg = [   ] g   63 000 g = [   ] kg   27 kg = [   ] g

STEP 2 (2.5 min) Rapid calculation

Start the timer

Convert these.

7 kg = [   ] g   90 000 g = [   ] kg   240 000 g = [   ] kg

6000 g = [   ] kg   30 kg = [   ] g   402 kg = [   ] g

10 kg = [   ] g   4 kg 80 g = [   ] g   7 kg 200 g = [   ] g

20 kg 3 g = [   ] g   6 kg 40 g = [   ] g

20 300 g = [   ] kg [   ] g   5002 g = [   ] kg [   ] g

403 029 g = [   ] kg [   ] g   72 050 g = [   ] kg [   ] g

STEP 3 (1.5 min) Challenge

Start the timer

Answer these.

8 kg + 60 g = [   ] g   7 kg + 400 g = [   ] g

3 kg − 200 g = [   ] g   12 kg + 80 g = [   ] g

90 kg − 750 g = [   ] g   2 kg − 5 g = [   ] g

4 kg + 50 000 g = [   ] kg   50 kg − 40 g = [   ] g

Time spent: _____ min _____ sec. Total: _____ out of 31   ©HarperCollins*Publishers* 2019

Date: _____

Day of Week: _____

STEP 1 (1 min) Warm-up

Start the timer

Convert these.

1 L = [ ] mL   5 L = [ ] mL   10 L = [ ] mL   200 L = [ ] mL

4000 mL = [ ] L   80 000 mL = [ ] L   96 000 mL = [ ] L   320 000 mL = [ ] L

STEP 2 (2.5 min) Rapid calculation

Start the timer

Convert these.

15 000 mL = [ ] L   3000 mL = [ ] L   30 L = [ ] mL   70 L = [ ] mL

900 000 mL = [ ] L   800 L = [ ] mL   18 000 mL = [ ] L   20 L = [ ] mL

70 000 mL = [ ] L   7 L 5 mL = [ ] mL   8 L 20 mL = [ ] mL   6 L 400 mL = [ ] mL

4090 mL = [ ] L [ ] mL

50 060 mL = [ ] L [ ] mL

72 003 mL = [ ] L [ ] mL

Mind Gym

Three customers each ordered one pancake from a restaurant.

In order to catch their train, they could not wait for more than 16 minutes.

The waiter said it would take at least 20 minutes to fulfil the order as the chef's only pan could only cook two pancakes at one time and it takes 5 minutes to bake each side of a pancake.

However, the head chef said he could fulfil the order in 15 minutes. How could he do this?

STEP 3 (1.5 min) Challenge

Start the timer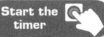

Answer these.

8 L + 5 mL = [ ] mL   8 L + 400 mL = [ ] mL

2 L – 70 mL = [ ] mL   7 L + 50 mL = [ ] mL

60 L – 240 mL = [ ] mL   4 L – 4 mL = [ ] mL

6 L + 60 000 mL = [ ] L   80 L – 20 000 mL = [ ] L

Date: _____

Day of Week: _____

STEP 1 (1 min) Warm-up

Start the timer

Answer these.

25 × 4 = ☐   360 ÷ 40 = ☐   120 × 80 = ☐   870 ÷ 3 = ☐

18 × 50 = ☐   42 × 300 = ☐   1200 ÷ 5 = ☐   98 ÷ 7 = ☐

STEP 2 (2.5 min) Rapid calculation

Start the timer

Complete the tables.

Speed	25 km/h	85 km/h	300 m/min		190 m/min
Time	3 h	8 h		20 s	
Distance			3600 m	400 m	570 m

Speed		2700 m/s		120 km/h	
Time	5 min	12 s	20 min		5 h
Distance	4800 m		3200 m	48 000 km	375 km

STEP 3 (1.5 min) Challenge

Start the timer

Answer these.

(20 × 9) ÷ 30 = ☐   (120 × 9) ÷ 60 = ☐

(840 ÷ 7) × 50 = ☐   (27 × 80) ÷ 4 = ☐

(125 × 8) ÷ 40 = ☐   (200 × 6) ÷ 40 = ☐

(1600 × 8) ÷ 20 = ☐   (327 × 8) ÷ 6 = ☐

Time spent: _____ min _____ sec. Total: _____ out of 26

Date: _____

Day of Week: _____

STEP 1 (1 min) Warm-up

Start the timer

Complete these.

1. The size of the space that an object takes up is called the .... of the object.

2. The edge length of a cube of volume $1\,cm^3$ is ....

3. The volume of a cube of edge length $1\,m$ is ....

4. The volume of a cuboid made from four cubes of edge length $1\,cm$ is ....

STEP 2 (2.5 min) Rapid calculation

Start the timer

Each cube is $1\,cm^3$. Fill in the missing numbers.

⬜ cubes; ⬜ cm^3

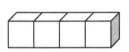

⬜ cubes; ⬜ cm^3

⬜ cubes; ⬜ cm^3

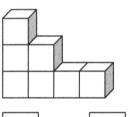

⬜ cubes; ⬜ cm^3

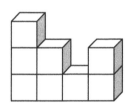

⬜ cubes; ⬜ cm^3

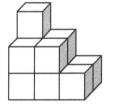

⬜ cubes; ⬜ cm^3

STEP 3 (1.5 min) Challenge

Start the timer

Each cube is $1\,cm^3$. Work out the volume of each shape.

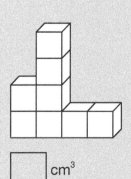

⬜ cm^3

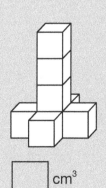

⬜ cm^3

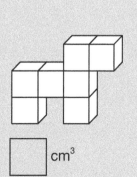

⬜ cm^3

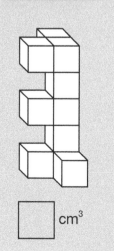

⬜ cm^3

Date: _____

Day of Week: _____

Start the timer

Convert these measures.

1 L = [   ] mL   5 L = [   ] mL   12 L = [   ] mL   20 L = [   ] mL

25 L = [   ] mL   30 mL = [   ] L   50 mL = [   ] L   66 mL = [   ] L

73 mL = [   ] L   100 mL = [   ] L

Start the timer

Convert these measures.

60 L = [   ] mL   0.9 L = [   ] mL   5.4 L = [   ] mL   0.3 L = [   ] mL

2019 mL = [   ] L   3 mL = [   ] L   2850 mL = [   ] L   8.3 L = [   ] mL

9.04 L = [   ] mL   82 mL = [   ] L   21.4 mL = [   ] L   600 mL = [   ] L

1230 mL = [   ] L   472 mL = [   ] L   49 mL = [   ] L   7.8 L = [   ] mL

3.05 L = [   ] mL   0.417 L = [   ] mL   16 L = [   ] mL   1.8 L = [   ] mL

Start the timer

 TIP *1 litre = 1000 millilitres = 1000 cubic centimetres (cm³)*

Convert these measures.

3 L 200 mL = [   ] mL   5 L 30 mL = [   ] mL   8006 cm³ = [   ] L

7010 cm³ = [   ] L   2 L 300 mL = [   ] L   30 L 50 mL = [   ] L

4002 cm³ = [   ] L   20.2 cm³ = [   ] L

15 010 cm³ = [   ] L   770 cm³ = [   ] L

Time spent: _____ min _____ sec. Total: _____ out of 40   ©HarperCollins*Publishers* 2019

Measurement of angles (1) **38**

STEP 1 (1 min) Warm-up

Choose the correct words or symbol from the box to complete each sentence.

| right ray vertex protractor degree side ∠ angle ° straight |

1. When two rays meet at a common point, they form an .... The common point is called

the .... of the angle. An angle can be represented with the symbol ....

2. An angle is measured using a .... The measurement of an angle is the ....

and it is represented with the symbol ....

3. At 6 o'clock, the angle formed by the hour hand and the minute hand on a clock face is a

.... angle. At 3 o'clock, the angle formed by the hour hand and the minute hand is a

.... angle.

STEP 2 (2.5 min) Rapid calculation

Measure the angles and complete the sentences.

1.
∠A = ☐° and is

a(n) .... angle.

A

2.

∠B = ☐° and is

a(n) .... angle.

B

3.

∠AOB = ☐° and is

a(n) .... angle.

4.
∠P = ☐° and is

a(n) .... angle.

P

5. N ——————— M   ∠MPN = ☐° and is a(n) .... angle.

P

STEP 3 (1.5 min) Challenge

Look at these angles:

45°   61°   89°   100°   12°   170°   90°   155°   30°   125°

Circle the acute angles, underline the obtuse angles and tick the right angle.

Date: _____

Day of Week: _____

Start the timer

Choose the correct word from the box to complete each sentence.

vertex	round	acute	right	central	straight	obtuse

An angle smaller than a right angle is called a(n) .... angle.

An angle greater than a right angle but smaller than a straight angle is called a(n) .... angle.

The angle formed by a ray rotating half a revolution about its endpoint is called a(n) .... angle.

The angle formed by a ray rotating $\frac{1}{4}$ revolution about its endpoint is called a(n) .... angle.

Start the timer

Measure the angles.

1.

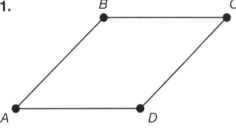

2.

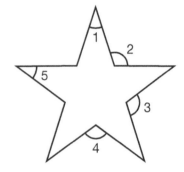

∠BAD = ☐° ∠ADC = ☐°

∠CBA = ☐° ∠BCD = ☐°

∠1 = ☐° ∠2 = ☐° ∠3 = ☐°

∠4 = ☐° ∠5 = ☐°

Start the timer

Thinking of a clock face, complete these sentences.

1. The hour hand moves through ☐ minute divisions per hour and the minute hand moves through

☐ minute divisions per hour.

2. At 2:45, the smaller angle formed by the hour hand and the minute hand is a(n) .... angle.

3. From 12 noon to 12 midnight, the hour hand moves through ☐°.

4. At ☐ and ☐ o'clock, the hour hand and the minute hand form a right angle.

5. From 12:10 to 12:20, the minute hand moves through ☐°.

Time spent: _____ min _____ sec. Total: _____ out of 20

STEP 1 (1 min) Warm-up

Start the timer

1. ∠E + ∠F = 180°. If ∠F = 95°, ∠E = ☐°.

2. ∠A − ∠B = 76°. If ∠A = 150°, ∠B = ☐°.

3. ∠D + ∠C = 90°. If ∠D = 40°, ∠C = ☐°.

4. ∠R − ∠S = 49°. If ∠S = 49°, ∠R = ☐°.

STEP 2 (2.5 min) Rapid calculation

Start the timer

1. If ∠1 = 25° and ∠2 = 40°, find ∠AOC.

∠AOC = ☐°

2. If ∠L = 65°, find ∠M.

∠M = ☐°

3. If ∠AOB = 40° and ∠AOB = ∠DOC, find ∠BOC.

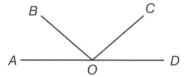

∠BOC = ☐°

4. If ∠AOC = 130°, find ∠BOC.

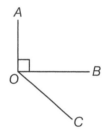

∠BOC = ☐°

5. If ∠1 = 75° and ∠2 = 40°, find ∠AOB.

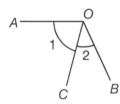

∠AOB = ☐°

6. If ∠AOC = 150°, find ∠BOC.

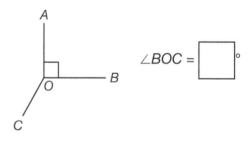

∠BOC = ☐°

STEP 3 (1.5 min) Challenge

Start the timer

1. If ∠AOB + 60° = a straight angle, then ∠AOB = ☐°.

2. If a round angle − ∠Y = 60°, then ∠Y = ☐°.

3. If ∠P = 60°, and ∠P is twice ∠Q, then the difference between ∠P and ∠Q is ☐°.

4. If ∠AOB + 103° = a straight angle, then ∠AOB = ☐°. If ∠AOB − ∠BOC = 40°, then ∠BOC = ☐°.

Time spent: _____ min _____ sec. Total: _____ out of 15

ANSWERS

Answers are given from top left, left to right, unless otherwise stated.

Test 1

Step 1:

9; 7; 0; 8

1; 3; 9; 0; 4

$4 \times 1000 + 2 \times 100 + 5 \times 10 + 0 \times 1$

$3 \times 10\,000 + 7 \times 1000 + 5 \times 100 + 2 \times 10 + 1 \times 1$

$3 \times 1000 + 5 \times 100 + 4 \times 10 + 9 \times 1$

Step 2:

Table completed from top: 50 201; 3292; 457 000; 80 040 630; 4 030 760; 5 003 000; 96 205 000; 7 900 800

Step 3:

First row: 20 402 000; twenty million, four hundred and two thousand

Second row: 3 050 006; three million, fifty thousand and six

Third row: 500 500 500; five hundred million, five hundred thousand and five hundred

Test 2

Step 1:

From top: 4872; 6408; 9001; 3029; 7770; 5700

Step 2:

From top: 7200; 4 927 005; 24 006 000; 36 420 000; 50 080 024; 260 000 000; 8 003 000 000; 1 504 080 000; 95 020 000; 106 000 660

Step 3:

<; >; <; <; <; <; <; >; >; <

Test 3

Step 1:

First row completed as: 5745; 5845; 5945; 6045

Second row completed as: 6976; 6876; 6776; 6676

Third row completed as: 46 863; 47 863; 48 863; 49 863

Fourth row completed as: 92 207; 91 207; 90 207; 89 207

Step 2:

First row completed as: 79 346; 80 346; 81 346; 82 346

Second row completed as: 94 079; 93 079; 92 079; 91 079

Third row completed as: 477 374; 487 374; 497 374; 507 374

Fourth row completed as: 220 458; 210 458; 200 458; 190 458

Fifth row completed as: 645 983; 745 983; 845 983; 945 983

Sixth row completed as: 809 921; 709 921; 609 921; 509 921

Step 3:

First row completed as: −10 000; 453 473; 443 473

Second row completed as: −100; 28 744; 28 644

Third row completed as: −100 000; 378 547; 278 547

Fourth row completed as: +1000; 361 975; 362 975

Test 4

Step 1:

498; 440; 800; 120; 3000; 24; 180; 7

Step 2:

300 000; 50 000; 59 940 000; 56 710 000; 1 500 000; 820 000; 3 500 000; 710 000; 900 000; 56 840 000; 740 000; 440 000; 7 410 000; 3 960 000; 50 040 000

Step 3:

First row: 4 500 000; 4 500 000

Second row: 7 370 000; 7 400 000

Third row: 7 950 000; 7 900 000

Fourth row: 3 190 000; 3 200 000

Test 5

Step 1:

1; 5; 20; 10; 100; 4; 6; 1000; 50; 500

Step 2:

3; 14; 300; 98; 29 7; 19; 600; 1600; 2010; 9; 150; 900; 65; 49

Step 3:

34; 95; 80; 1100; 1500; 1900; 1970; 1995

Test 6

Step 1:

1. positive, negative   **2.** can, cannot

3. positive, negative (in either order)

4. positive, negative (in either order)

Step 2:

Positive numbers: +24; 5.9; +10.6; 32.5; $+\frac{4}{7}$

Negative numbers: −19; −2.08; $-\frac{2}{25}$; −57

Step 3:

107; 96; 106; 97.6; 103.5; 106.3; 94.2

Test 7

Step 1:

From the top: origin; right; left

Step 2:

1. $A = -2.5$; $B = 0$; $C = 2.5$; $D = -4$; $E = 1$

2.

3. right; 6   **4.** −3.5

Step 3:

1. left; 5.5   **2.** +7 or −7   **3.** −4 or +2

4. 9; 4

5. +2   **6.** 1, 2, 3

Answers

Test 8

Step 1:

15 867; 10 476; 14 858; 7251

Step 2:

54 107; 132 270; 85 095; 116 433; 26 879; 151 699; 137 408; 168 361

Step 3:

132 101; 244 334; 157 419; 230 323

Test 9

Step 1:

1228; 3822; 3055; 1410

Step 2:

1903; 24 288; 27 669; 31 042; 37 958; 17 428

Step 3:

$$\begin{array}{r} 94928 \\ - 59309 \\ \hline 35619 \end{array} \qquad \begin{array}{r} 59078 \\ - 27063 \\ \hline 32015 \end{array} \qquad \begin{array}{r} 84984 \\ - 14230 \\ \hline 70754 \end{array}$$

$$\begin{array}{r} 62458 \\ - 28754 \\ \hline 33704 \end{array} \qquad \begin{array}{r} 51025 \\ - 45657 \\ \hline 5368 \end{array} \qquad \begin{array}{r} 28207 \\ - 20977 \\ \hline 7230 \end{array}$$

Test 10

Step 1:

23; 2.5; 19.3; 11.5; 96; 10; 1.7; 0.24; 1.82; 4.19

Step 2:

8.1; 38.47; 92.3; 36; 7800; 8.6; 1.293; 0.46; 1.93; 0.099; 795.6; 0.175; 4.23; 350; 0.9; 0.356

Step 3:

0.076; 380; 0.55; 99; 3.45; 2.67; 0.6; 3569

Test 11

Step 1:

560; 35; 23.4; 20; 80.2; 3.9; 0.029; 4.1; 0.0025; 0.0106

Step 2:

768; 62.9; 0.0259; 3180; 47.08; 0.8; 0.1216; 290; 2.601; 3050; 0.0704; 3280; 0.084; 0.59; 0.044

Step 3:

100; 10; 100; 1000; 100; 10; 1000; 100

Test 12

Step 1:

1200; 700; 480; 760; 900; 960; 910; 2000

Step 2:

1700; 1500; 9000; 1480; 786; 990; 3600; 2000; 1200; 1000; 930; 750; 3800; 6600; 775

Step 3:

924; 418; 624; 957; 480; 396; 4225; 275

Test 13

Step 1:

19; 13; 11; 12; 29; 10; 28; 23; 12; 14

Step 2:

4; 5; 3; 3; 50; 50; 30; 40; 4; 50; 3; 60; 60; 40; 3

Step 3:

90; 20; 144; 720; 31; 1000; 210; 30; 56; 1260

Test 14

Step 1:

From top:

1. 5, 5, 50   **2.** 2, 2, 20

3. 9, 9, 90   **4.** 9, 9, 90

Step 2:

400; 60; 80; 300; 90; 60; 40; 50; 50; 50; 70; 60; 70; 70; 50

Step 3:

40; 150; 1600; 135; 99; 780; 205; 120; 60; 128

Test 15

Step 1:

15; 16; 12; 15; 18; 21; 13; 18; 16; 16

Step 2:

16; 14; 13; 17; 18; 12; 47; 21; 25; 4; 5; 16; 19; 35; 27

Step 3:

32; 183; 39; 152; 99; 54; 33; 8

Test 16

Step 1:

1. 1, 2, 4, 8, 16   **2.** 1, 2, 3, 4, 6, 8, 12, 24

3. 1, 2, 7, 14, 49, 98   **4.** 32, 64, 96

5. 18, 36, 54, 72, 90

Step 2:

1. 1, 2, 4, 8   **2.** 1, 5   **3.** 1, 3, 9   **4.** 1, 2, 4

5. 12, 24, 36, 48, 60   **6.** 24, 48, 72, 96, 120

7. 15, 30, 45, 60, 75   **8.** 18, 36, 54, 72, 90

For 5–8, other suitable answers are possible.

Step 3:

1. 4; 6; 12; 16

2. 16; 91; 24; 36

Test 17

Step 1:

81; 25; 9; 16; 64; 343; 64; 1; 8; 27

Answers

Step 2:

121; 125; 169; 216; 324; 1000; 144; 512; 225; 2; 361; 14; 400; 8; 256

Step 3:

8; 1; 13; 11; 7; $3^2 \times 5^2$; $2^2 \times 7^2$; $2^2 \times 9^2$ or $3^2 \times 6^2$

Test 18

Step 1:

1. 72, 48, 30, 62, 80, 60, 58   **2.** 30, 15, 80, 45, 60

Step 2:

90; 78; 50; 86; 225; 195; 125; 215; 188; 164; 152; 116; 470; 410; 380; 290

Step 3:

37; 154; 420; 243; 58; 191; 329; 332; 149; 199

Test 19

Step 1:

1. 2, 23, 31, 97, 67   **2.** 72, 9, 15, 49, 56, 63

Step 2:

$3 \times 3 \times 5$;
$2 \times 2 \times 2 \times 2 \times 2$;
$2 \times 3 \times 11$;
$3 \times 3 \times 3$;
$2 \times 2 \times 2 \times 2 \times 3$;
$3 \times 3 \times 3 \times 3$;
$2 \times 3 \times 3 \times 3$;
$2 \times 2 \times 3 \times 7$;
$2 \times 2 \times 2 \times 2 \times 2 \times 3$

Step 3:

$2 \times 2 \times 17$;
$2 \times 2 \times 23$;
$2 \times 2 \times 2 \times 11$;
$2 \times 3 \times 13$;
$2 \times 2 \times 2 \times 3 \times 3$;
5×17;
$2 \times 3 \times 5$;
3×23

Test 20

Step 1:

99; 12; 137; 2440; 2000; 132; 330; 570; 1300

Step 2:

2. $356 \div 4 \times (470 - 362) = 9612$

3. $3870 \div [(238 - 195) \times 9] = 10$

4. $(1360 - 247) \div (18 + 35) = 21$

5. $(105 - 4) \times (49 + 33) = 8282$

6. $(3045 - 128 \times 15) \div 45 = 25$

Other suitable answers are possible.

Step 3:

1. $650 - 15 \times 15 = 425$ **2.** $450 \div (50 + 50 \times 2) = 3$

3. $280 \div (16 + 54) = 4$ **4.** $(49 - 38) \times (64 \div 8) = 88$

Mind Gym:

9, 5, 7

Test 21

Step 1:

Description of rate	Rate	Unit rate
60 mm of rain over 4 days	$\dfrac{60\,\text{mm of rain}}{4\ \text{days}}$	15 mm of rain per day
6 books for £30	$\dfrac{£30}{6\ \text{books}}$	£5 per book
120 seeds in 3 rows	$\dfrac{120\ \text{seeds}}{3\ \text{rows}}$	40 seeds per row
125 oranges in 25 bowls	$\dfrac{125\ \text{oranges}}{25\ \text{bowls}}$	5 oranges per bowl
132 hours of work in 12 days	$\dfrac{132\ \text{hours}}{12\ \text{days}}$	11 hours per day

Step 2:

Description of rate	Rate	Unit rate	Calculation
420 points in 6 games	$\dfrac{420\ \text{points}}{6\ \text{games}}$	70 points per game	630 points in 9 games
280 pages in 8 days	$\dfrac{280\ \text{pages}}{8\ \text{days}}$	35 pages per day	385 pages in 11 days
£252 for 7 hours work	$\dfrac{£252}{7\ \text{hours}}$	£36 per hour	£540 for 15 hours work
297 girls in 9 groups	$\dfrac{297\ \text{girls}}{9\ \text{groups}}$	33 girls per group	495 girls in 15 groups

Step 3:

Description of rate	Rate	Unit rate	Calculation
760 kilometres on 8 litres of fuel	760 km / 8 L	95 km / L	1235 km on 13 L of fuel
438 points in 6 plays of a video game	438 points / 6 plays	73 points / play	876 points in 12 plays
£756 for 9 tickets	£756 / 9 tickets	£84 per ticket	£1260 for 15 tickets
720 kg for 12 crates	720 kg / 12 crates	60 kg per crate	1080 kg for 18 crates

Test 22

Step 1:

117; 900; 440; 360; 1360; 311; 4800; 7575; 790

Step 2:

1625; 1200; 50; 350; 1717; 808; 6; 198; 1400; 36; 12; 8000; 96; 396; 4851

Step 3:

140; 198; 3650; 5; 2700; 4000; 125; 40

Test 23

Answers are given in their lowest terms. Equivalent answers are possible.

Step 1:

$\frac{15}{17}$, $\frac{2}{3}$, $\frac{17}{20}$, $\frac{7}{8}$, $\frac{5}{19}$, $\frac{9}{14}$, $\frac{4}{9}$, $\frac{2}{11}$

Step 2:

$\frac{8}{9}$, $\frac{84}{107}$, $\frac{5}{11}$, $\frac{16}{25}$, $\frac{1}{8}$, $\frac{19}{90}$, $\frac{20}{23}$, $\frac{1}{5}$, $\frac{25}{27}$, $\frac{38}{53}$, $\frac{30}{217}$, $\frac{37}{49}$, $\frac{5}{9}$, $\frac{1}{8}$, $\frac{100}{113}$, $\frac{43}{169}$

Step 3:

$\frac{69}{100}$, $\frac{5}{9}$, $\frac{2}{7}$, $\frac{9}{223}$, $\frac{16}{17}$, $\frac{45}{119}$, $\frac{34}{55}$, $\frac{54}{89}$

Test 24

Step 1:

>; <; >; <; >; <; >; <; >; <

Step 2:

<; >; <; <; >; >; >; >; <; <; >; >; <; <; <; >

Step 3:

$\frac{1}{3} > \frac{1}{4} > \frac{1}{7}$; $\frac{85}{99} > \frac{73}{99} > \frac{37}{99}$; $\frac{3}{5} > \frac{1}{5} > \frac{1}{15}$; $\frac{8}{9} > \frac{9}{9} > \frac{3}{13}$;

$\frac{13}{14} > \frac{13}{20} > \frac{4}{20}$; $\frac{7}{10} > \frac{3}{10} > \frac{3}{11}$

Test 25

Step 1:

1. $23\frac{1}{4}$, $1\frac{6}{15}$, $40\frac{2}{3}$, $72\frac{3}{8}$

2. $\frac{9}{4}$, $\frac{25}{24}$, $\frac{19}{8}$, $\frac{8}{7}$　　　3. $\frac{7}{13}$, $\frac{3}{8}$, $\frac{84}{100}$, $\frac{181}{366}$

Step 2:

1. $4\frac{4}{7}$; $9\frac{1}{3}$; $6\frac{3}{4}$; $8\frac{3}{4}$; $5\frac{7}{12}$; $18\frac{1}{2}$

2. $\frac{17}{7}$, $\frac{19}{6}$, $\frac{68}{15}$, $\frac{151}{12}$, $\frac{143}{34}$, $\frac{149}{14}$

Step 3:

1. $\frac{7}{3}$, $\frac{13}{8}$, $\frac{61}{9}$, $\frac{107}{34}$　　　2. $4\frac{3}{7}$; $6\frac{5}{8}$; 9; $10\frac{1}{4}$

Test 26

Answers are given as mixed numbers in their lowest terms. Equivalent answers are possible.

Step 1:

$\frac{4}{5}$; 1; $\frac{11}{15}$, $\frac{13}{24}$; 1; $\frac{4}{15}$; $\frac{4}{9}$, $\frac{9}{23}$, $\frac{17}{37}$, $\frac{31}{102}$

Step 2:

$\frac{2}{3}$, $\frac{2}{5}$, $\frac{1}{2}$; $\frac{11}{20}$; $\frac{7}{23}$, $\frac{31}{45}$; $\frac{5}{9}$; $\frac{5}{9}$, $\frac{7}{18}$, $\frac{13}{32}$, $\frac{79}{106}$, $\frac{29}{46}$

Step 3:

$\frac{53}{70}$; 1; $\frac{1}{6}$; $\frac{11}{135}$; $\frac{1}{16}$; $\frac{31}{34}$; $1\frac{1}{8}$; $\frac{2}{9}$

Test 27

Answers are given as mixed numbers in their lowest terms. Equivalent answers are possible.

Step 1:

$\frac{6}{7}$; $\frac{16}{19}$, $\frac{10}{27}$; 1; $\frac{9}{10}$, $\frac{4}{5}$; $1\frac{2}{3}$; $\frac{3}{8}$; $1\frac{3}{13}$

Step 2:

$\frac{9}{13}$, $\frac{3}{5}$; $1\frac{2}{3}$; $\frac{36}{73}$, $\frac{43}{60}$; $1\frac{1}{15}$; $4\frac{13}{14}$; $\frac{3}{4}$; $2\frac{1}{10}$; $3\frac{3}{5}$; $8\frac{1}{2}$; $20\frac{8}{9}$; $12\frac{9}{14}$;

$108\frac{12}{17}$

Step 3:

$48\frac{36}{95}$; $21\frac{77}{100}$; $15\frac{5}{9}$; $65\frac{39}{47}$; $61\frac{1}{9}$; $108\frac{27}{28}$; 49; $81\frac{9}{56}$

Test 28

Step 1:

1. 0.864 < 2.468 < 24.68 < 25.068 < 25.860

2. 6.032 > 5.766 > 5.667 > 5.660 > 0.320

Step 2:

0.5; 30.202; 40; 1.48; 1.9; 400; 3.94; 8; 0.081; 7.08; 2.91; 51.01

Step 3:

1. 23.70; 67.00; 12.50; 4.50; 87.10; 31.00

2. 45.900; 78.560; 9.120; 92.000; 1.700; 37.000

Test 29

Step 1:

Box 1: 3.90, 1.4000, 5.600

Box 2: 3.90, 1.4000, 5.600, 201.00, 3.0200, 50.30

Box 3: 10.005, 4.01, 1000, 10.001, 204.09

Step 2:

3.04; 7.05; 40.03; 8.09; 100.2; 9; 50.505; 3.02; 0.3; 3.03; 6.2; 5.8; 10.1; 0.75; 3.99; 20.2

Step 3:

0.36; 0.04; 1.10; 1.06; 3.04; 0.90; 46.35; 70.03

Test 30

Step 1:

2.6; 11.3; 8.5; 11.8; 17.9; 2; 6.9; 3.5; 9; 6.1

Step 2:

11.4; 1.72; 12.5; 1.7; 3.94; 9.9; 11.09; 2.37; 2.22; 7.4; 1.04; 10.5; 43.72; 4.4; 0.35; 8.05

Step 3:

2.4; 12.7; 13.3; 7.9; 1.3; 10.9; 1.8; 8

Test 31

Step 1:

0.3; 2.6; 2.5; 1.4; 8.15; 0.05; 0.4; 8.7; 0.34; 3.8

Step 2:

0.3; 2.6; 0.51; 0.1; 8.36; 0.95; 0.7; 0.05; 3.4; 8.91; 47.3; 1.2; 10.8; 3.6; 0.67; 9.75

Step 3:

5.3; 0.6; 7; 9.96; 17; 3.52; 9.1; 3.3

Test 32

Step 1:

23; 51; 66; 97; 49; 88; 3; 91; 9; 24; 71

Step 2:

Percentage	7%	77%	19%	30%	59%	90%	1%	43%	99%	5%
Fraction	$\frac{7}{100}$	$\frac{77}{100}$	$\frac{19}{100}$	$\frac{30}{100}$	$\frac{59}{100}$	$\frac{90}{100}$	$\frac{1}{100}$	$\frac{43}{100}$	$\frac{99}{100}$	$\frac{5}{100}$
Decimal	0.07	0.77	0.19	0.3	0.59	0.9	0.01	0.43	0.99	0.05

Answers

Step 3:

Percentage	3%	62%	80%	11%	33%	8%	44%	85%	60%	13%
Fraction	$\frac{3}{100}$	$\frac{62}{100}$	$\frac{80}{100}$	$\frac{11}{100}$	$\frac{33}{100}$	$\frac{8}{100}$	$\frac{44}{100}$	$\frac{85}{100}$	$\frac{60}{100}$	$\frac{13}{100}$
Decimal	0.03	0.62	0.8	0.11	0.33	0.08	0.44	0.85	0.6	0.13

Test 33
Step 1:
1000; 4000; 1; 8; 5; 80 000; 63; 27 000
Step 2:
7000; 90; 240; 6; 30 000; 402 000; 10 000; 4080; 7200; 20 003; 6040; 20; 300; 5, 2; 403, 29; 72, 50
Step 3:
8060; 7400; 2800; 12 080; 89 250; 1995; 54; 49 960

Test 34
Step 1:
1000; 5000; 10 000; 200 000; 4; 80; 96; 320
Step 2:
15; 3; 30 000; 70 000; 900; 800 000; 18; 20 000; 70; 7005; 8020; 6400; 4, 90; 50, 60; 72, 3
Step 3:
8005; 8400; 1930; 7050; 59 760; 3996; 66; 60
Mind Gym:
Bake the first sides of Pancake A and Pancake B. Then bake the second side of Pancake A and the first side of Pancake C. Finally, bake the second sides of Pancakes B and C. It would take 15 minutes in all.

Test 35
Step 1:
100; 9; 9600; 290; 900; 12 600; 240; 14
Step 2:
First table: 75 km; 680 km; 12 min; 20 m/s; 3 min
Second table: 960 m/min; 32 400 m; 160 m/min; 400 h; 75 km/h
Step 3:
6; 18; 6000; 540; 25; 30; 640; 436

Test 36
Step 1:
1. volume 2. 1 cm 3. 1 m³ 4. 4 cm³
Step 2:
4, 4; 4, 4; 4, 4; 7, 7; 8, 8; 11, 11
Step 3:
8; 8; 7; 10

Test 37
Step 1:
1000; 5000; 12 000; 20 000; 25 000; 0.03; 0.05; 0.066; 0.073; 0.1

Step 2:
60 000; 900; 5400; 300; 2.019; 0.003; 2.85; 8300; 9040; 0.082; 0.0214; 0.6; 1.23; 0.472; 0.049; 7800; 3050; 417; 16 000; 1800
Step 3:
3200; 5030; 8.006; 7.01; 2.3; 30.05; 4.002; 0.0202; 15.01; 0.77

Test 38
Step 1:
1. angle, vertex, ∠
2. protractor, degree, °
3. straight, right
Step 2:
1. 110°, obtuse 2. 50°, acute 3. 50°, acute
4. 90°, right 5. 180°, straight
Step 3:
Circled: 45°, 61°, 89°, 12°, 30°
Underlined: 100°, 170°, 155°, 125°
Ticked: 90°

Test 39
Step 1:
From top:
acute; obtuse; straight; right
Step 2:
1. $\angle BAD = 45°$, $\angle ADC = 135°$, $\angle CBA = 135°$, $\angle BCD = 45°$
2. $\angle 1 = 36°$, $\angle 2 = 108°$, $\angle 3 = 108°$, $\angle 4 = 108°$, $\angle 5 = 36°$
Step 3:
1. 5, 60 2. obtuse 3. 360
4. 3, 9 5. 60

Test 40
Step 1:
1. 85 2. 74 3. 50 4. 98
Step 2:
1. (25 + 40 =) 65 2. (180 − 65 =) 115
3. (180 − 80 =) 100 4. (130 − 90 =) 40
5. (75 + 40 =) 115 6. (360 − 90 − 150 =) 120
Step 3:
1. 120 2. 300 3. 30 4. 77, 37